1st Grade Science Volume 1

© 2013 Todd Deluca
OnBoard Academics, Inc
Newburyport, MA 01950

800-596-3175
www.onboardacademics.com

Table of Contents

Day and Night

Day and Night

We all know that it's dark at night and light during the day. You may have heard that this is because the sun rises every day and sets or goes down every night making it dark. That's not exactly correct, let's take a look at what really causes night and day on earth.

The Spinning Earth

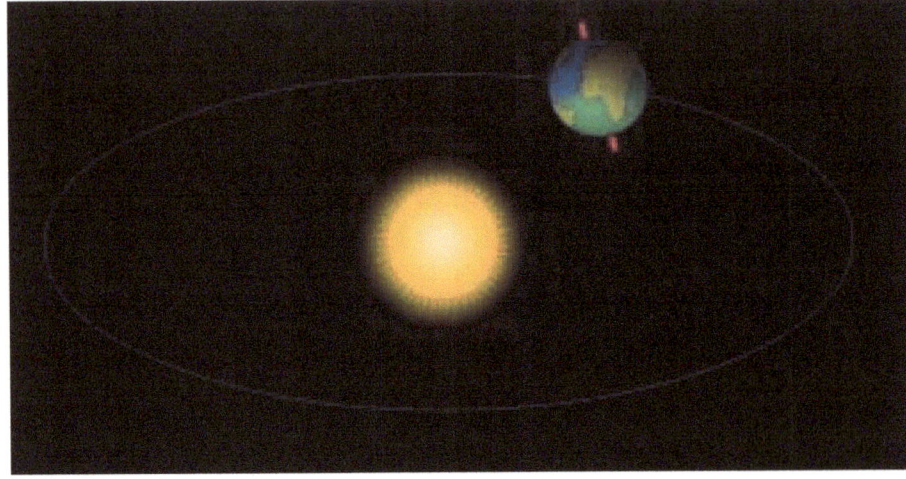

We are looking at a picture of the earth that has a pattern that it follows around the sun. The earth also spins on its own axis as it rotates around the sun. As the sun spins on its axis the part of the earth that faces the sun is having day and the part of the earth facing away is having night. It takes the earth 24 hours to make one entire revolution on its axis. That is why one day is 24 hours long. For most of us, half the day it is light outside and half the day it is dark outside because we are either in the sun's light or we are not.

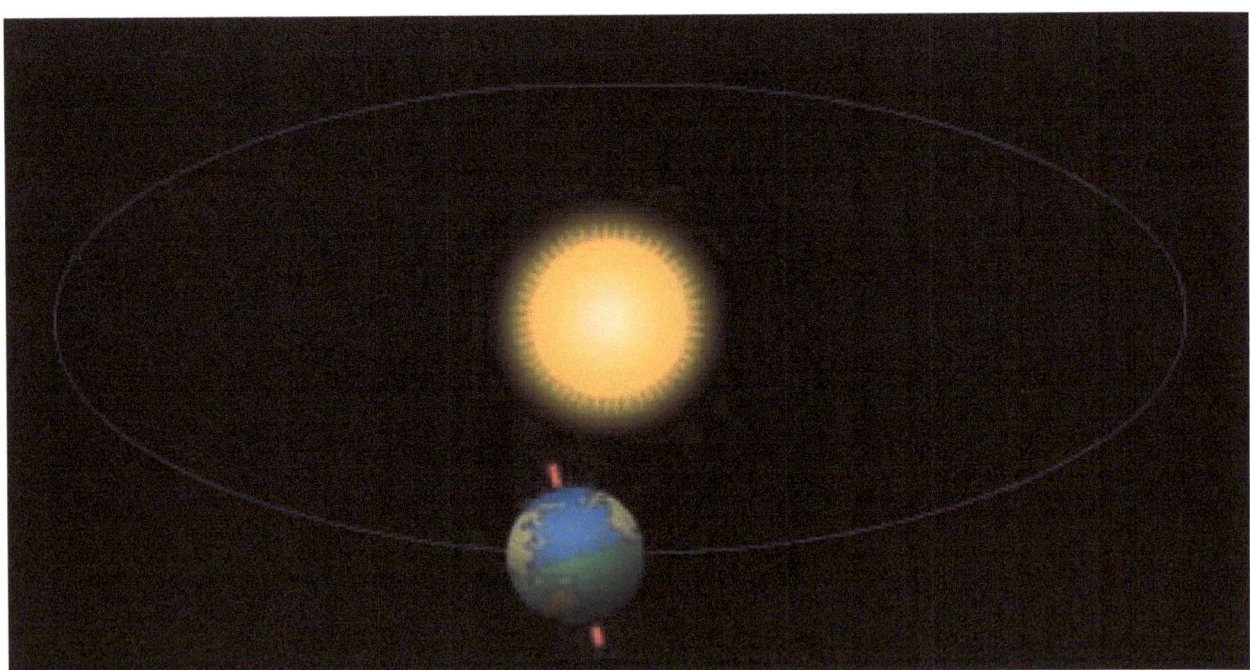

Can you see how the rotation causes night and day?

North America is marked with a red flag. Label if it is night or day in each illustration.

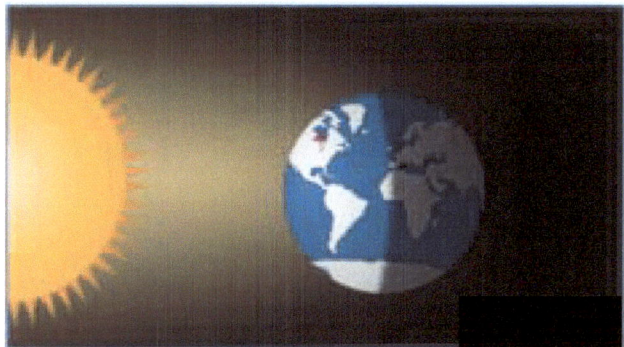

Night or Day?

_____

Night or Day?

_____

As you can see, each location that you mark on Earth experiences day or night depending on when it is facing the Sun. That's because Earth spins on an imaginary axis. It takes 24 hours (one day) for Earth to spin all the way around.

Match the times of day and night with the images.
Write your answers in the space provided.

12 a.m.	6 a.m.	12 p.m.	6 p.m.
midnight	sunrise	noon	sunset

Why do shadows change during the day? _____

_____

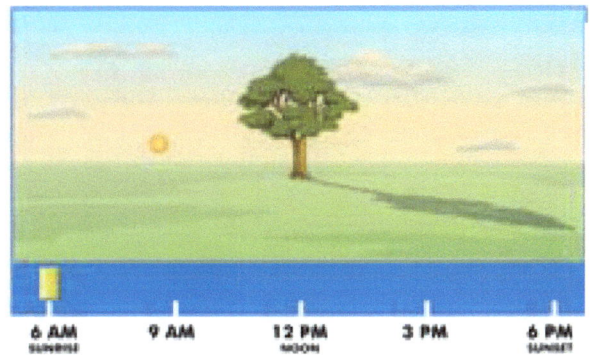

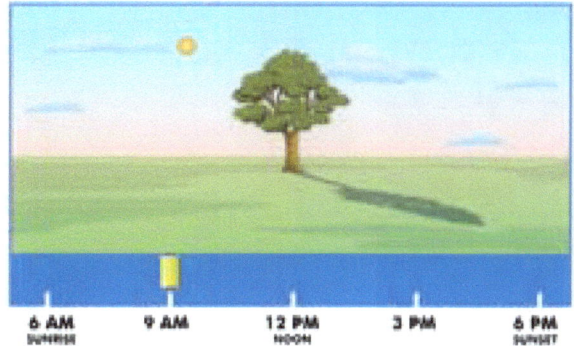

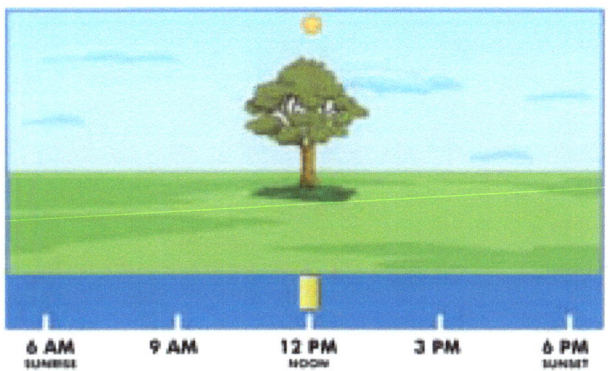

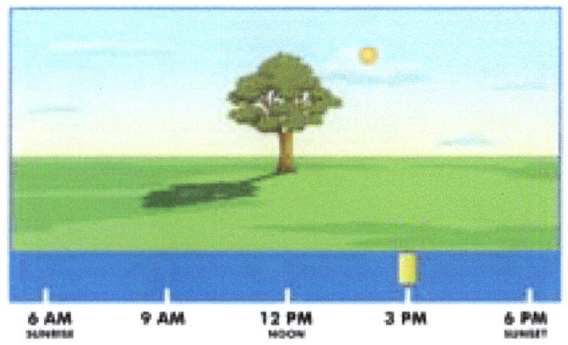

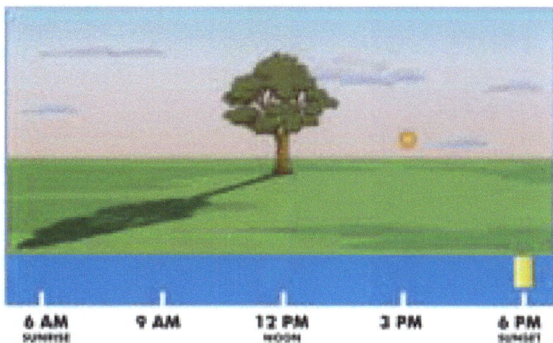

Shadows are formed when objects block light from the Sun (or light from another source). Shadows change according to the position of the Sun as it appears to move from East to West in the sky. For example, at noon, the Sun is directly overhead and so this makes a shadow appear to be very short.

Connect the time of day with the proper shadow.

| 6 a.m. | 9 a.m. | 12 noon | 3 p.m. | 6 p.m. |

Name: _____

Day and Night Quiz

1. Because of the movement of the Earth, the Sun appears to rise in the _____.
    a. East
    b. West
    c. North
    d. South

2. You can know time by following the shadows the Sun casts on the ground. True or false?

3. The shadows cast by the Sun at noon are _____>
    a. Long
    b. Short
    c. Towards the right

4. We have night and day on Earth because of the Earth's revolution around the Sun. True or false?

5. The portion of the Earth facing the Sun has night. True or false?

6. The Earth takes _____ hours to rotate on its axis.

7. The Earth takes _____ days to rotate around the Sun.

Patterns in the Night Sky

What picture do these stars form. Look at the stars and select and answer from the choices provided. Write your answer in the box.

bull   spoon   swan   bicycle

Hint

If you connect the stars with a line, it looks like a spoon or a ladle. In the old days, a ladle was called a dipper and so we call this star pattern the Dipper or the Big Dipper. In other cultures, this star pattern is also called the Plough and the Bear.

Constellations Form Pictures in the Sky

A constellations is the name we give to a group of stars that seem to form a picture in the night star. Constellations have been observed for centuries and have been used to try to explain the universe and for story time and myths.

Constellations were also used as maps in the sky for navigation at night. This is because although you can see different stars in winter and summer all the stars that you can see at night are always in the same place when compared to each other.

Which constellation do you think this is?

Connect the stars. If you don't see a recognizable figure(s) try again. See the hint below before you get started.

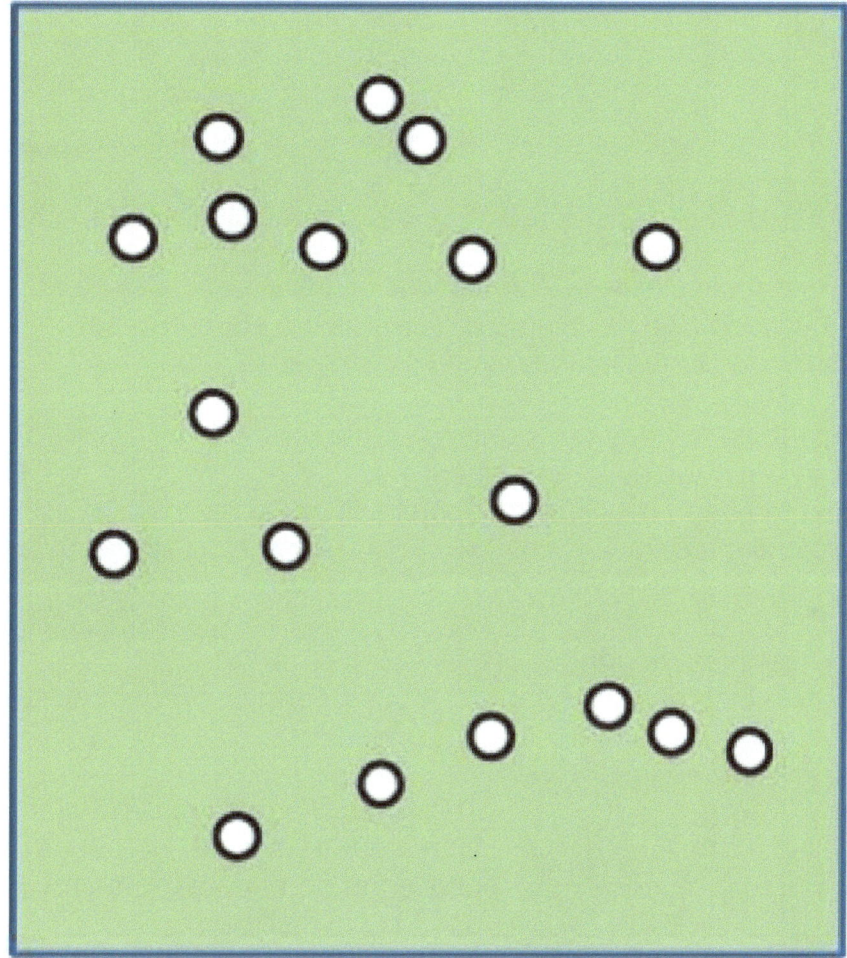

This is the constellation called Gemini which people thought resembled two twin brothers holding hands.

The North Star

If you patiently observe the night sky you will notice that the stars seem to spin in a

 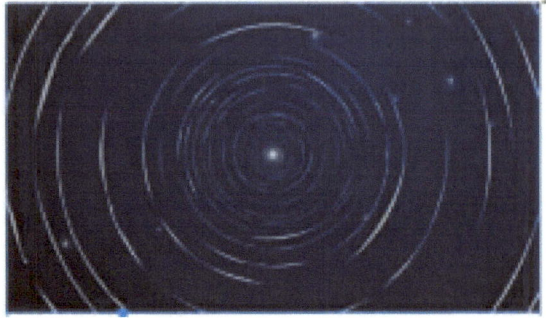

circle around one particular star. The special star that seems to be at the center of all this attend is the North Star. While it looks like all the stars are moving in a circle around the North Star, in reality they aren't moving at all .

The reason they look like they are spinning is that the earth is spinning like a top on an imaginarily axis. We don't notice that the earth is spinning because all of the air around us is spinning too.

Lets look at the earth as though we are far away on a space ship. Notice that there are stars all around earth in every direction.

The North Star also called Polaris is directly above our north pole. Because of its position above the north pole it is a very important star used in navigation.

The North star is the end star in the constellation known as the little dipper.

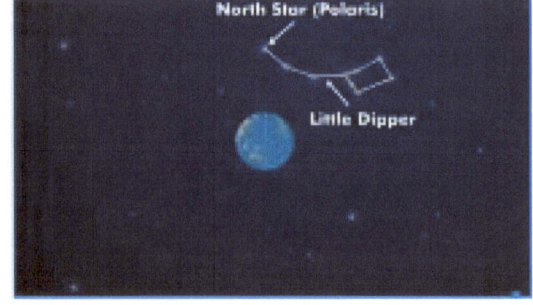

You can also use the big dipper to find the north star.

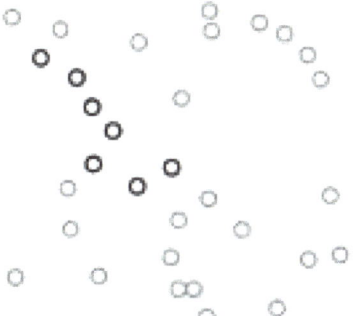

First, you have to find the Big Dipper. It is always visible in the Northern Hemisphere because it is made of very bright stars.

Then find the two stars at the end of the Dipper. These two stars are called the Pointer Stars.

Draw an imaginary line that goes straight through the Pointer Stars and which is about five times as long as the Pointer Stars are apart. Your line will hit the North Star.

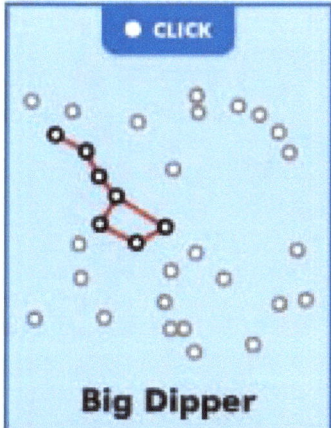

Big Dipper

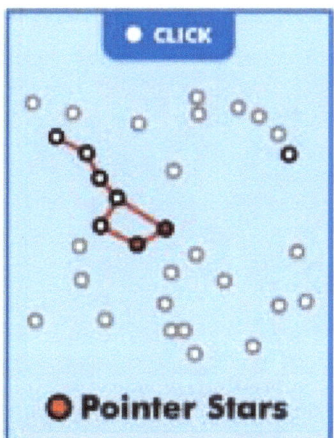

● **Pointer Stars**

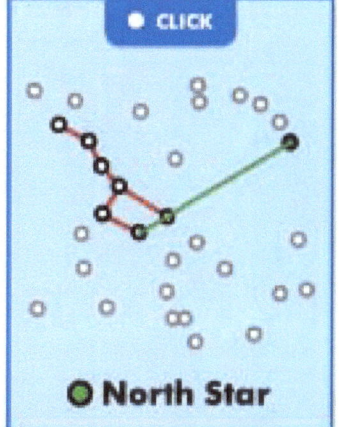

● **North Star**

This strategy always works because the stars are always in the same position relative to each other.

Circle the North Star in each of these skies.

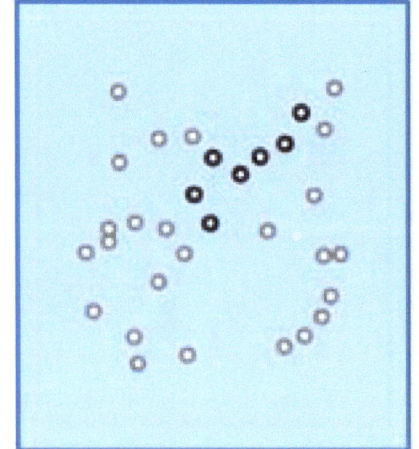 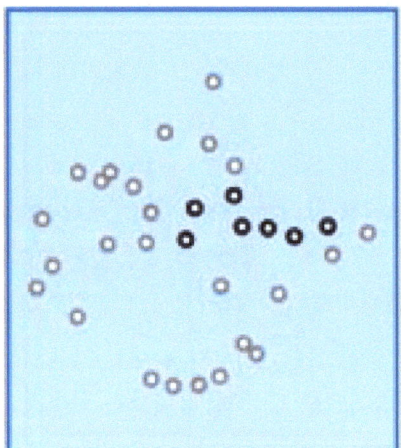 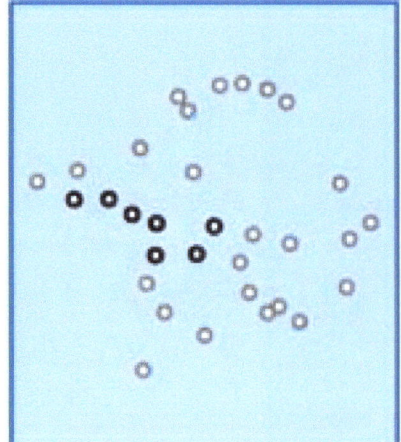

Remember, the stars appear to spin in a circle at night because Earth is spinning. So, the Big Dipper sometimes will look upside-down or sideways.

Identify these famous constellations and label them them with the suggestions provided. Here is a hint; Taurus is a bull and Orion is a hunter.

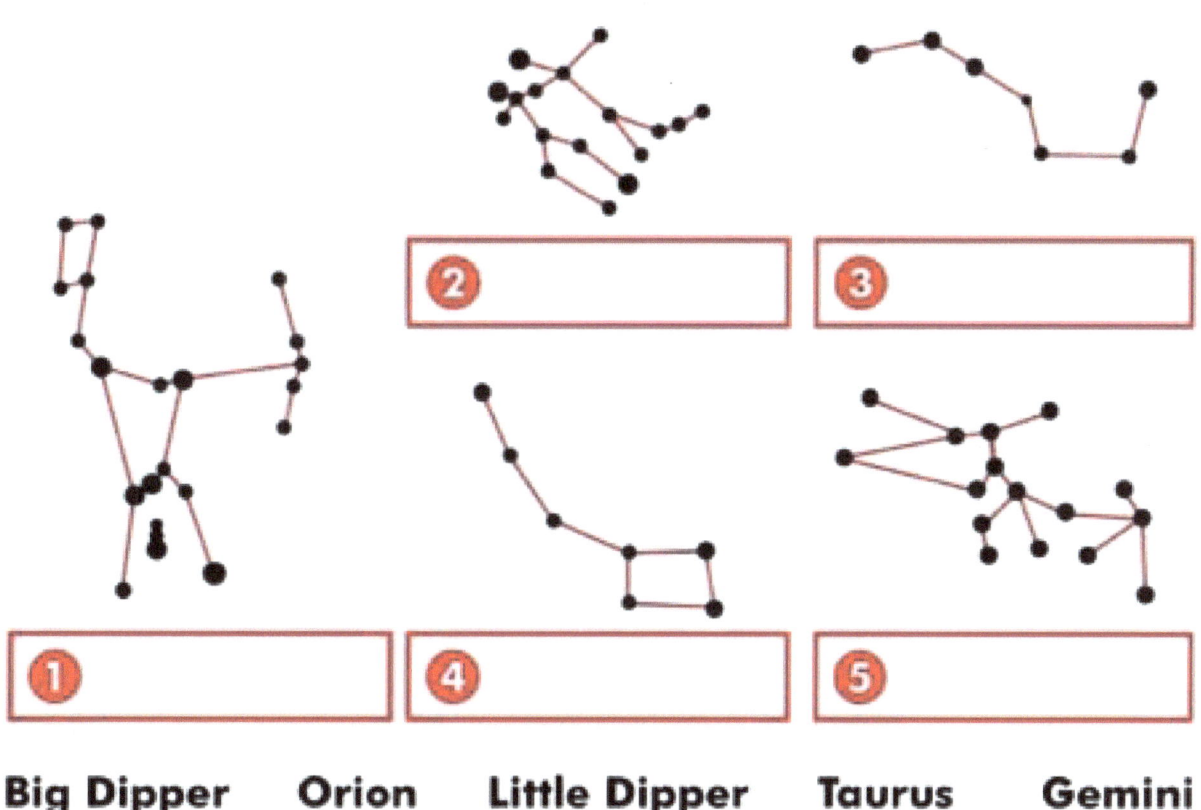

Big Dipper   Orion   Little Dipper   Taurus   Gemini

Invent you own constellation by connecting some stars, making a shape and naming your constellation.

Constellation Name: _____

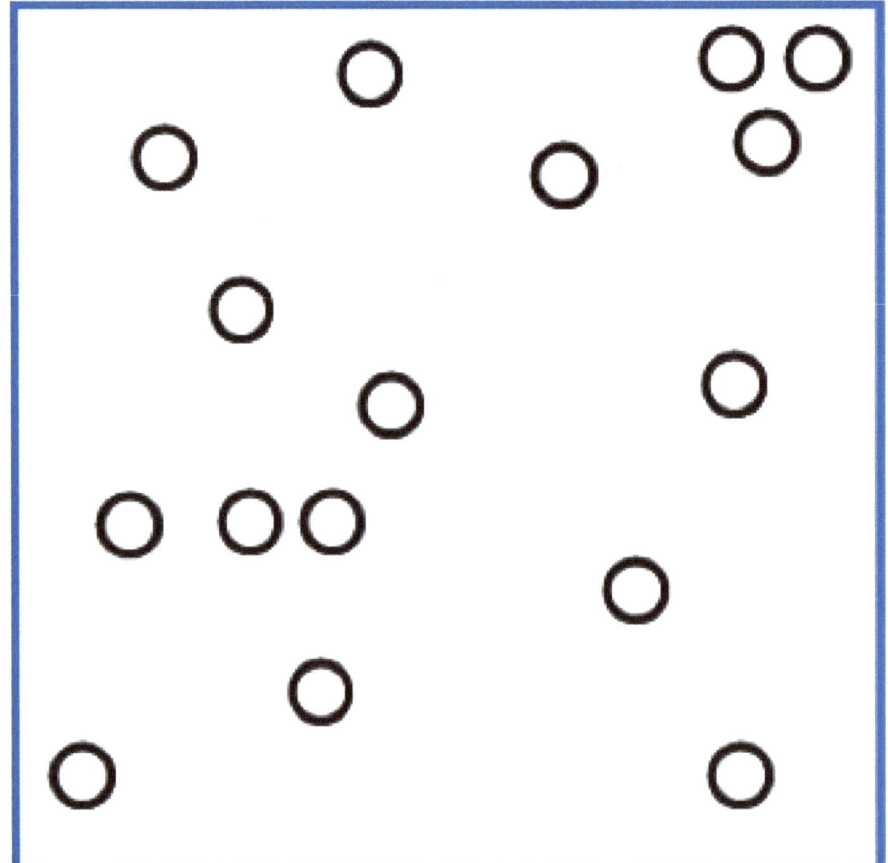

Name: _____

Patterns in the Night Sky Quiz

1. A star pattern that looks like a ladle or spoon is called the Big Dipper. True or false?

2. A group of stars forming a picture in the night sky is called a _____.
    a. constellation
    b. planet
    c. heavenly body

3. Constellations were used as _____ in the sky for navigation at night.
    a. maps
    b. symbols
    c. legends

4. Stars seem to spin in a circle around a special star called the _____?
    a. East Star
    b. South Star
    c. North Star

5. Stars seem to be moving because the Earth is spinning like a top on an imaginary axis. True or false?

Sound

Sounds You Hear

Look at these pictures.

 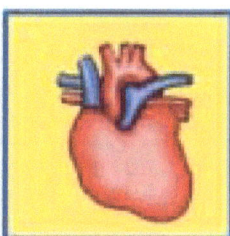   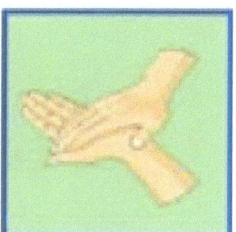

Can you fill in the box with the sound that might be made by each illustration?

What causes sound?

When a guitar is played and a sound is made, what happens? _____

_____

_____

When a drum is used and a sound is made, what happens. _____

_____

_____

Sounds are made by vibrations.

The guitar is silent

The guitar is playing

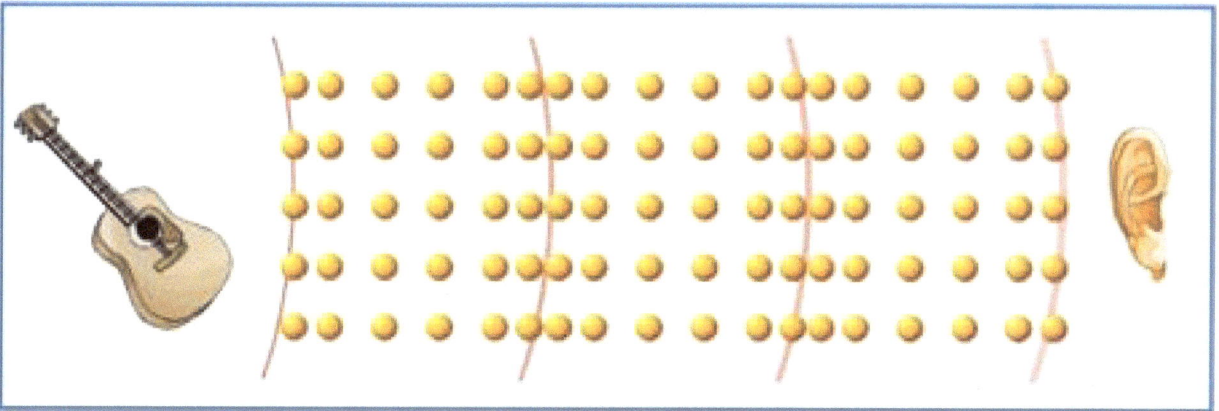

When an object vibrates, it causes vibrations in the air around it. These vibrations very quickly reach your ear and cause your eardrum (a soft thin soft part inside your ear) to vibrate. Signals are then sent to your brain and you hear sound.

Put a √ next to the objects that vibrate when they make a sound.

Can you arrange these objects by their sounds being softest to loudest? Draw the item in the correct box.

Objects that make loud sounds vibrate more than objects that make soft sounds. Loud sounds have more energy and so their vibrations travel further.

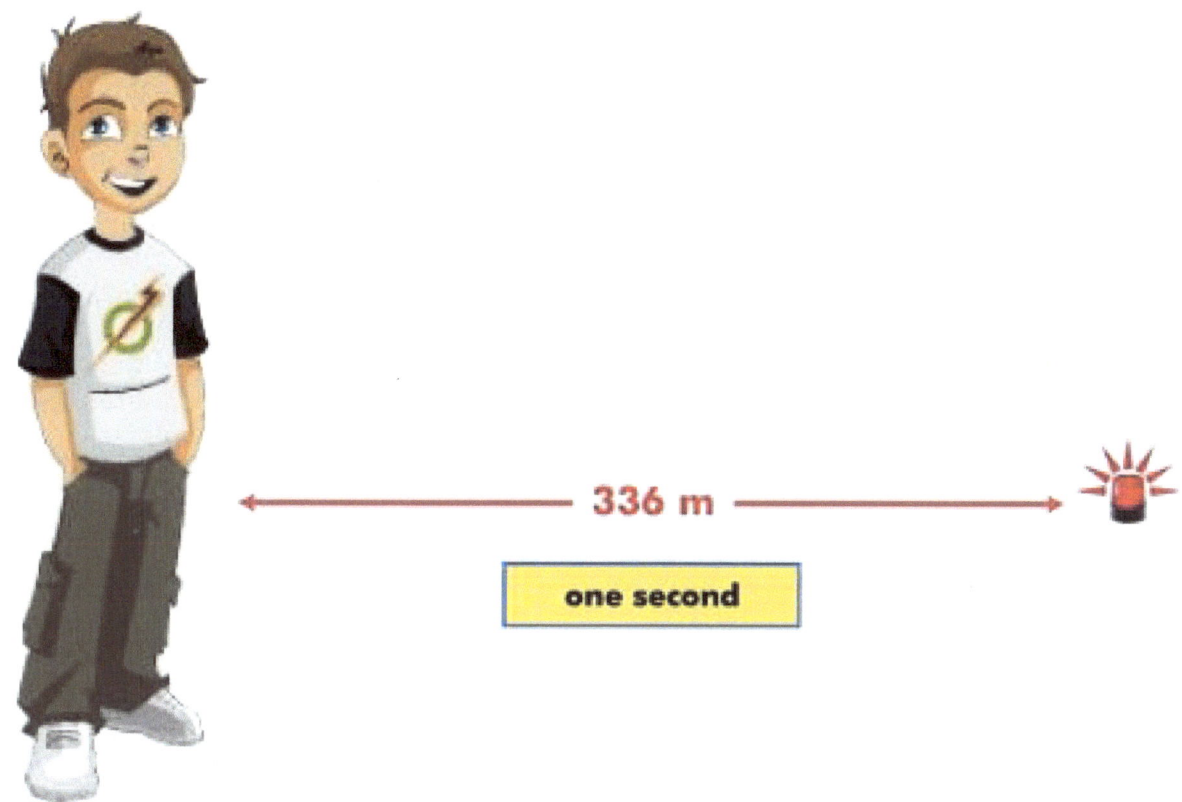

Owen is standing 336 meters(that's about one quarter of a mile) from a siren when it goes off. After the Owen sees the light from the siren, he won't hear the sound for one second.

The speed of sound changes depending on what material it's traveling through.

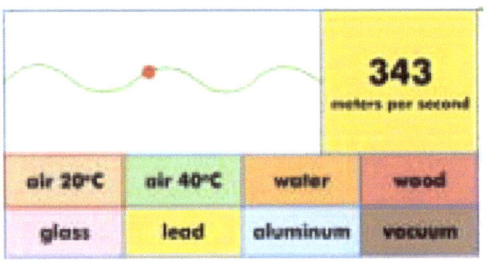

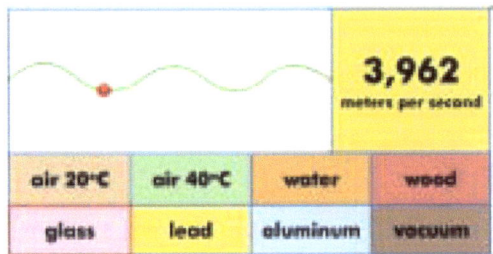

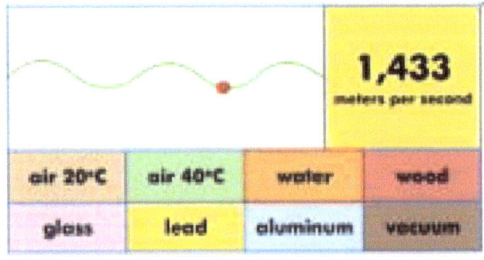

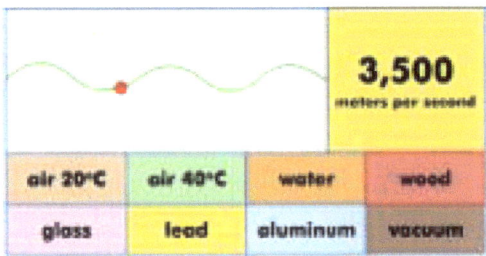

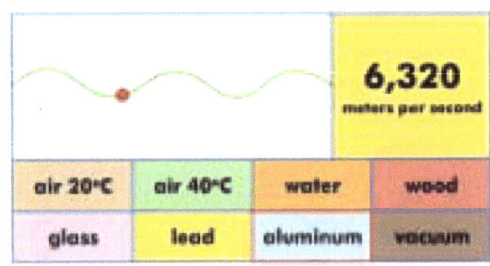

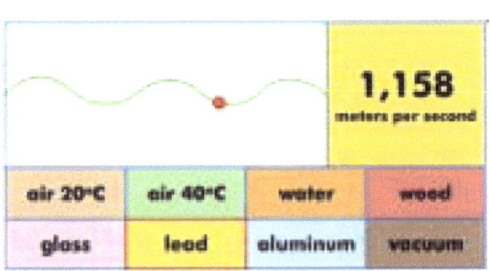

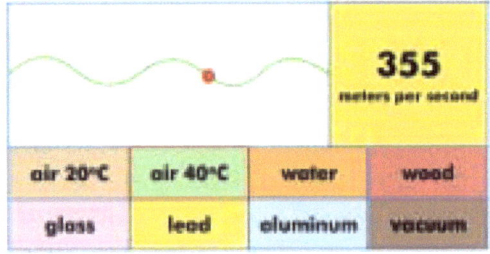

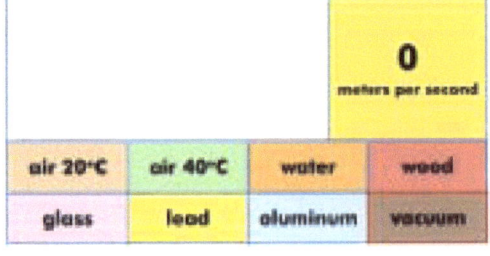

In which substance does sound travel the fastest? _____

In which substance does sound travel the slowest? _____

In which substance does sound not travel at all? _____

www.onboardacademics.com

Sound Quiz

1. Sound is a form of energy produced due to vibration. True or false?

2. Which of these objects make a sound; pencil, boiling water, book? _____

3. Vibration is when something moves back and forth quickly. True or false?

4. Which of the following is the softest sound?
   a. Phone ringing
   b. Butterfly fluttering
   c. Car honking

5. Apart from air, sounds can also travel through other materials. True or false?

6. Which does sound travel through faster, air or water?

   _____

7. Which is the loudest of the following sounds?
   a. Drums beating
   b. Leaves rustling
   c. Birds chirping

Properties of Light

Do you know how long it takes light to travel from the sun to the earth?

Light travels amazingly fast. A beam of light could travel around the Earth over seven times in about a second. However, because the Sun is so far away, it actually takes sunlight about eight minutes to reach Earth.

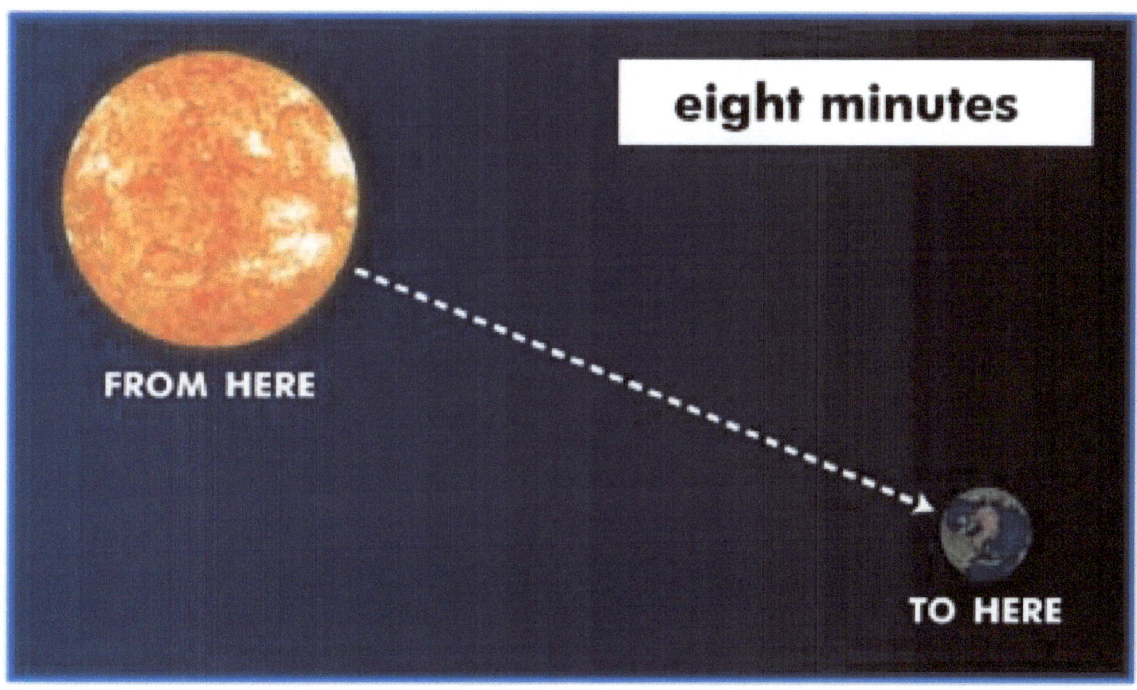

eight minutes

FROM HERE

TO HERE

The Source of Light

All light has a source and travels in a direction. If you can see light, then the light must have started somewhere and traveled to get there.

Circle the source of the light.

Which direction will the light travel?

Draw light beams from each object in the direction that they will travel.

When traveling light is blocked it makes a shadow.

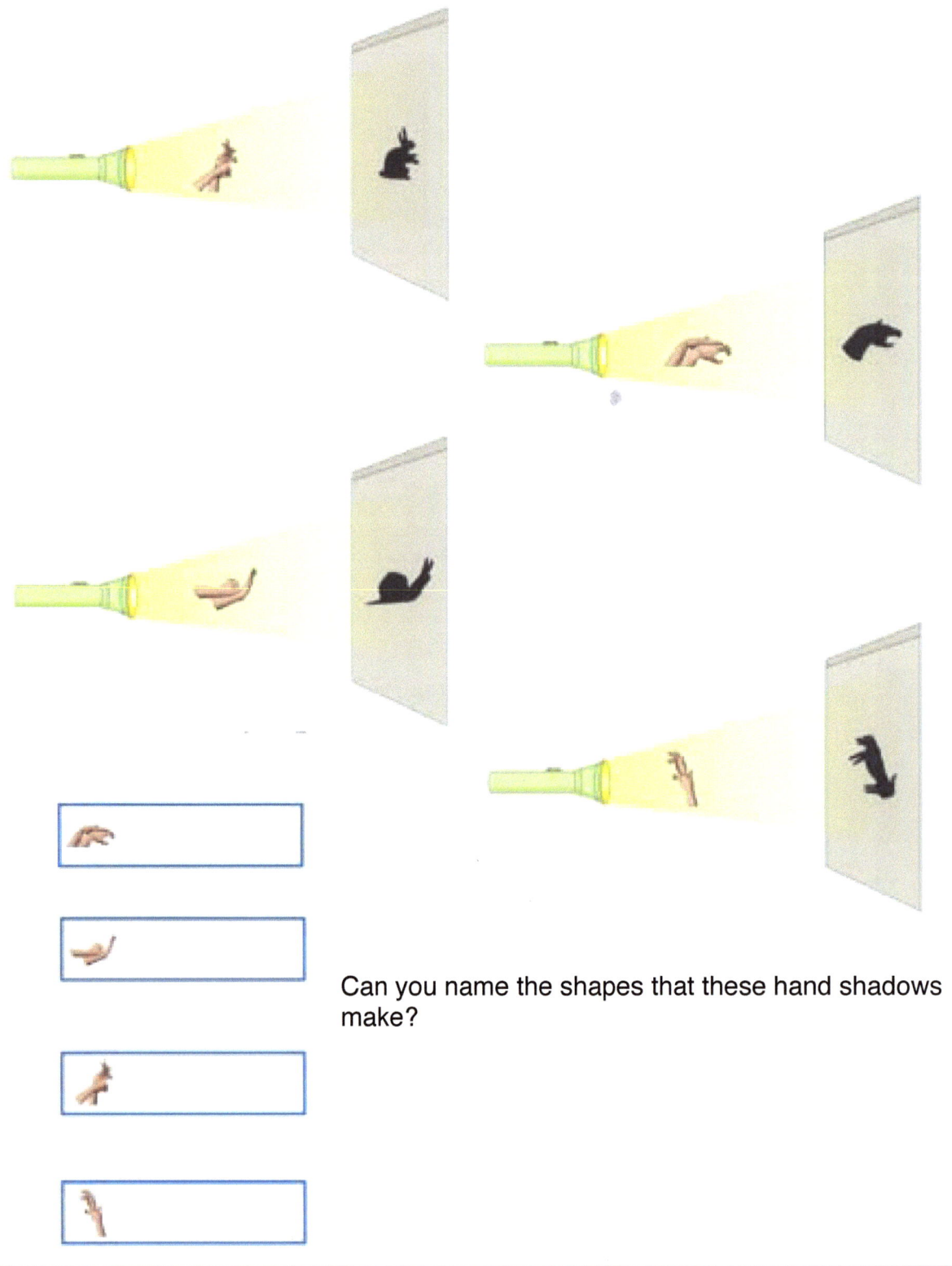

Can you name the shapes that these hand shadows make?

Light bounces or reflects off of objects.

When light travels from a light source and hits an object it bounces or reflects off of the object in many different directions. Some of the reflective light that bounces from the

object hits your eye entering through the pupil and then travels to a part of the eye called the retina. The retina sends a signal to the brain that processes the information and gives us the sensation of sight.

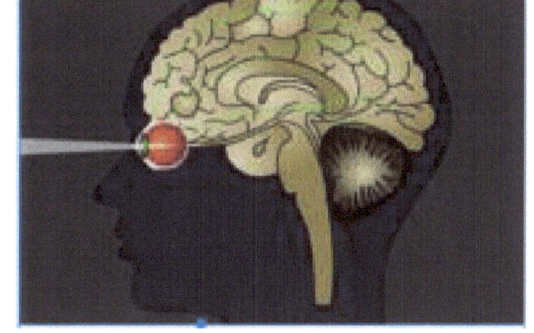

You only see the flower when light is reflected from the flower and hits your eye. If there is no light the flower is still there but you can't see it because there is no light to reflect off of the flower and into your eye.

Reflection or bouncing light is why you can see yourself in a mirror. When you see yourself in the mirror , first the light reflects off of you and then reflects off of you and

onto a mirror. Because mirrors are made with a very reflective material that makes almost all light reflect off of them, the light reflects off of the mirror and back into your eye. This is why you see your own image in the mirror.

So what it a reflection?

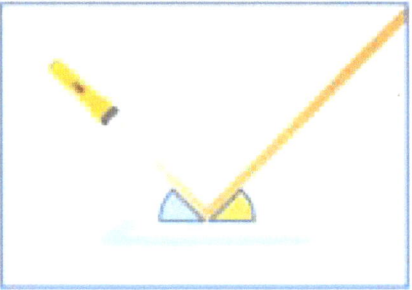

When light hits a shiny surface like a mirror, it gets reflected and bounces off in a new direction.

A mirror can be used to change the direction of a beam of light.

Sort these objects by how reflective they are.

Very reflective	Not very reflective

Properties of Light Quiz

Fill in the blanks.

All light has a _____ ; a place where the light comes from. Light doesn't just appear, it _____ in a direction. Light travels very _____. In fact, _____ is faster than light. If something blocks light, a _____ is created. When light hits an object, the light is _____ off the object. If the light reflects into your _____, then you can see the object. Some materials, like mirrors, are very _____.

absorbs	slowly	disappears	shadow
nothing	source	reflective	brain
quickly	travels	reflected	eyes

Observations with Properties

What are the five senses?

Unscramble to solve.

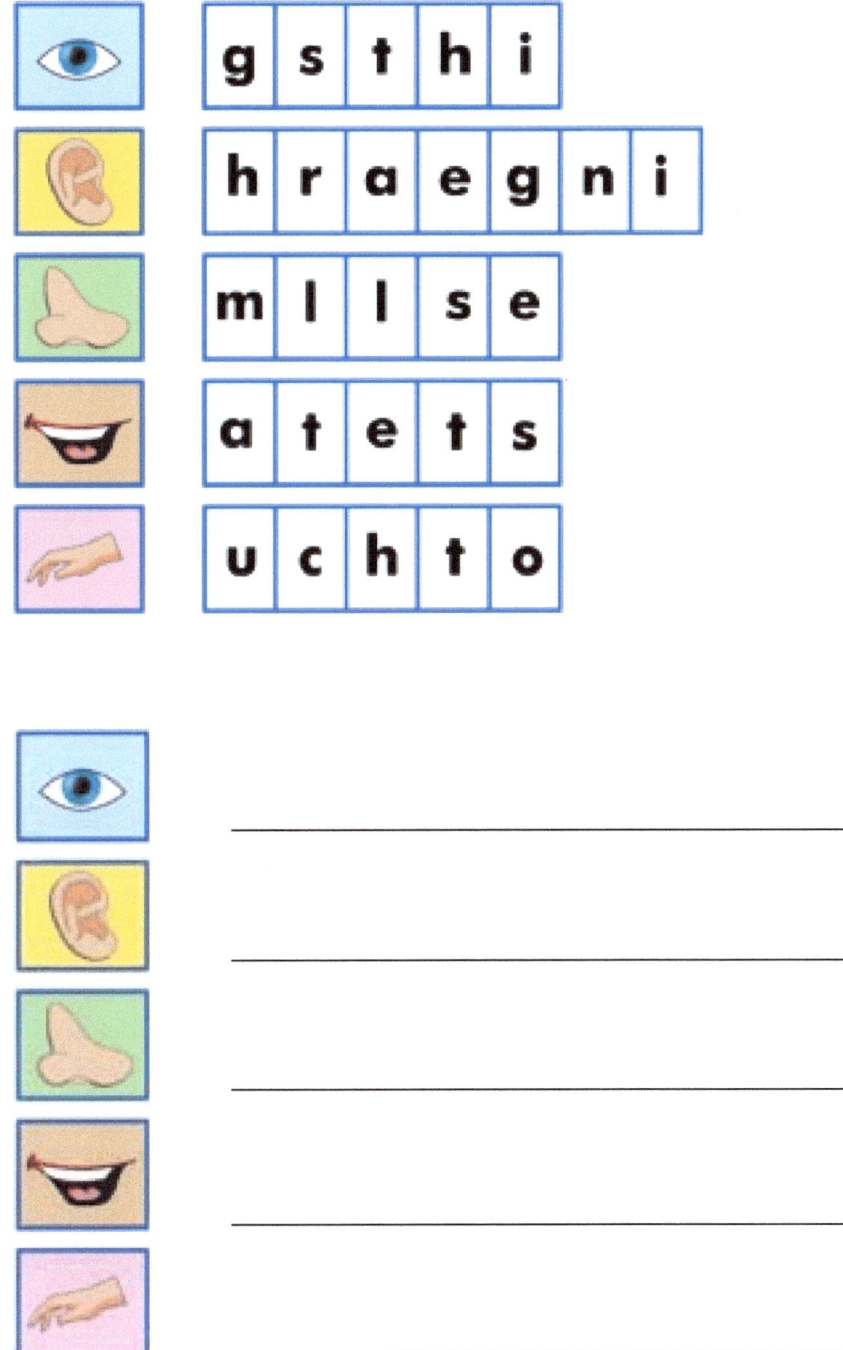

Label the five sense organs for humans.

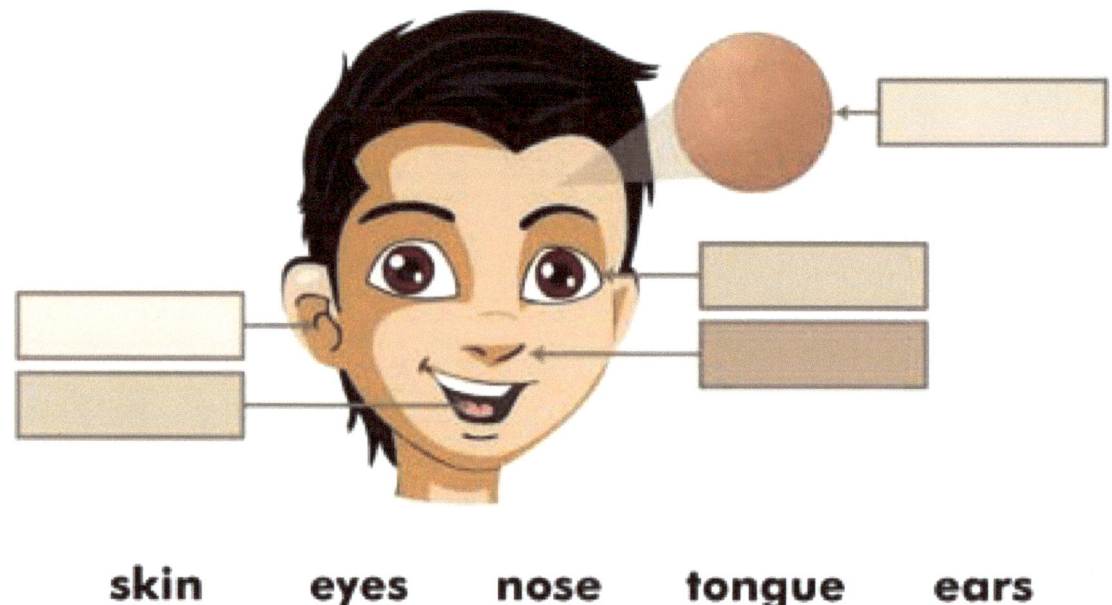

skin   eyes   nose   tongue   ears

Humans have five sense organs: eyes, ears, nose, tongue, and skin. These organs send messages to the brain which give us the sensation of sight, sound, smell, taste, and touch.

Match the senses and the sensory organs.

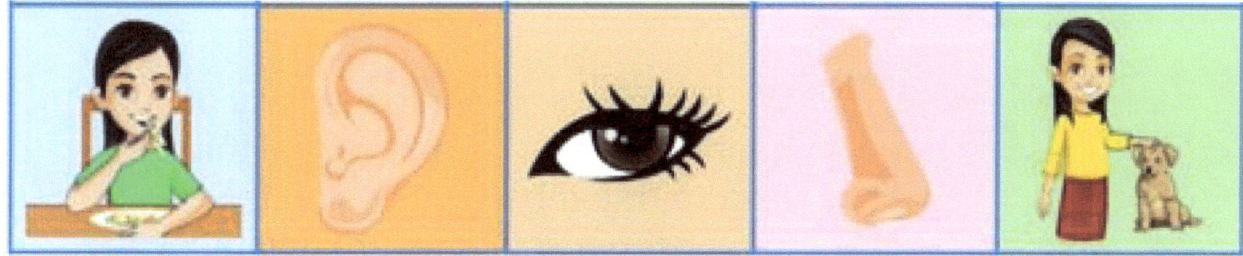

How we use our senses to make observations.

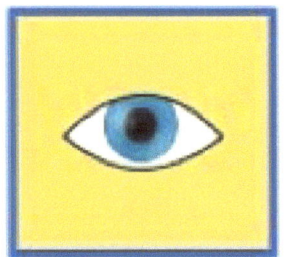

We use our eyes to observe the visual properties of an object such as its shape color or size.

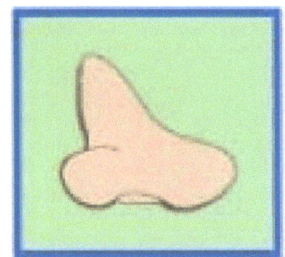

We use our nose to observe an object's odor; what it smells like. We may be able to identify an object by its odor alone. Can you identify a skunk without seeing it?

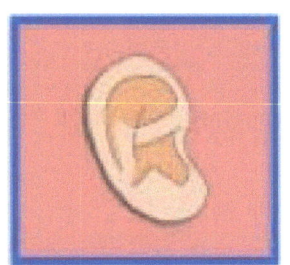

We use our eyes to make observations about a sound an object makes. Is the sound loud or soft, pleasant or unpleasant, from a hollow or solid object?

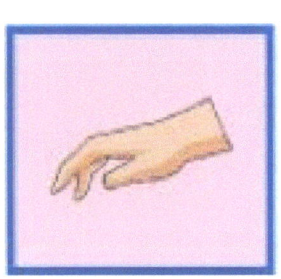

We use our skin, and especially our hands, to touch objects to feel their texture or temperature.Using touch we can observe soft, smooth, bumpy or cold as just a few examples.

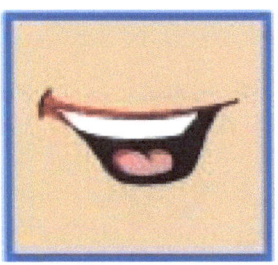

We use our tongue to taste an object, usually food. We can observe if it tastes pleasant, salty, sweet, sour or bitter. Only test taste when you a sure the object is safe to eat.

Which senses am I using to observe these properties?
Match the sense with the likely comment.

Ouch! That's hot!

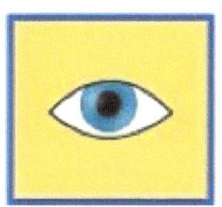

The stone was green and shiny

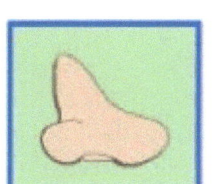

Yuck! That milk is sour.

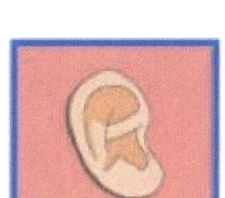

I didn't touch it but I knew the wood was damp due to the musky odor.

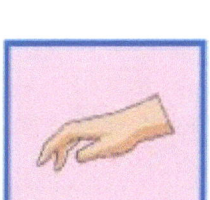

I had a jagged dorsal fin and that's how I knew it was a great white shark.

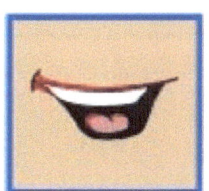

Ratata-tat. I think there is a woodpecker nearby.

What's in the bag?

Using the clues for the five senses identify what is in the bag.

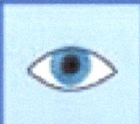 I can't see it because it's in the bag, but I see that it fits in a bag.

 It doesn't have a strong odor but the smell reminds me of autumn.

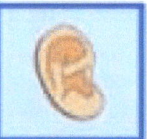 It doesn't have any sound.

 Its hard and sort of roundish. It has a tiny stick coming from one end and both ends are indented a bit.

 Its crunch and juicy and has a pleasant fruity taste. It's not sour at all.

Now let's see if we can describe something for someone else. Think of an object and write the clues next to each sense describing our the sense would evaluate the object is the bag. Ask a friend or teacher to try and guess your object in the bag.

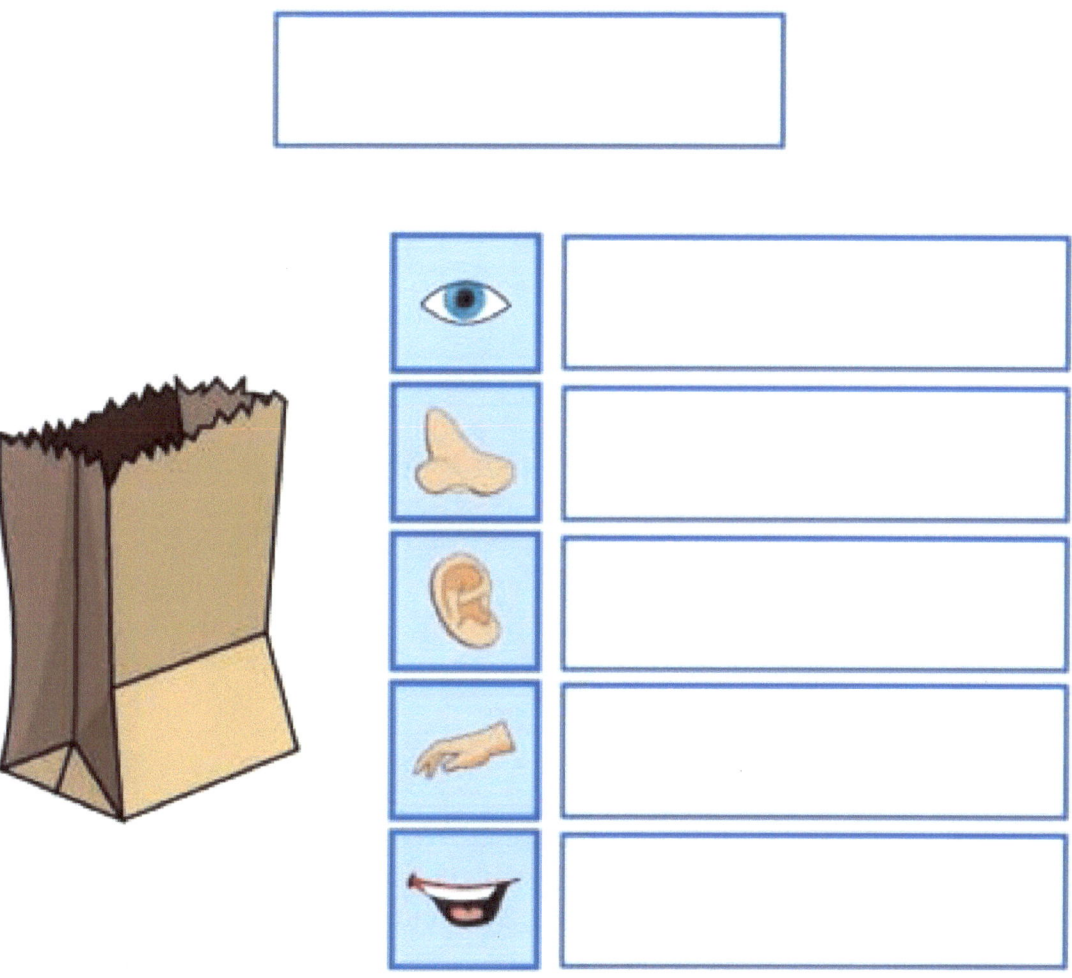

Using tools to make observations about properties.

Connect the tool with the correct description.

We use this tool to measure time.

We use this tool to measure an object's size.

  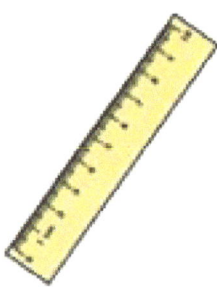  

We use this tool to help us to see tiny objects or objects in great detail.

We use this tool to measure weight.

We use this tool to measure temperature.

Observations with Properties Quiz

1. Anything that we observe about a material with our senses is called a _____ of that material.
    a. shape
    b. color
    c. property
    d. smell

2. Which of the following is NOT a property of a material?
    a. color
    b. red
    c. shape
    d. softness

3. The nose is a sensory organ. True or false?

4. Different senses help us identify different properties of a material. True or false?

5. We _____ cars honking.
    a. see
    b. smell
    c. hear
    d. feel

Physical Properties of Materials

What would you make these objects with?

Write your answer in the box provided.

paper   wood   cloth   rubber   metal   plastic

Materials are chosen for different tasks because of their **physical properties**. Physical properties describe important characteristics of a material such as its strength, flexibility, transparency, conductivity, waterproofness, or magnetism. Let's explore what these properties mean and which materials have which properties.

Match the physical proper ties with their definition.

conductive

magnetic

flexible

strong

waterproof

transparent

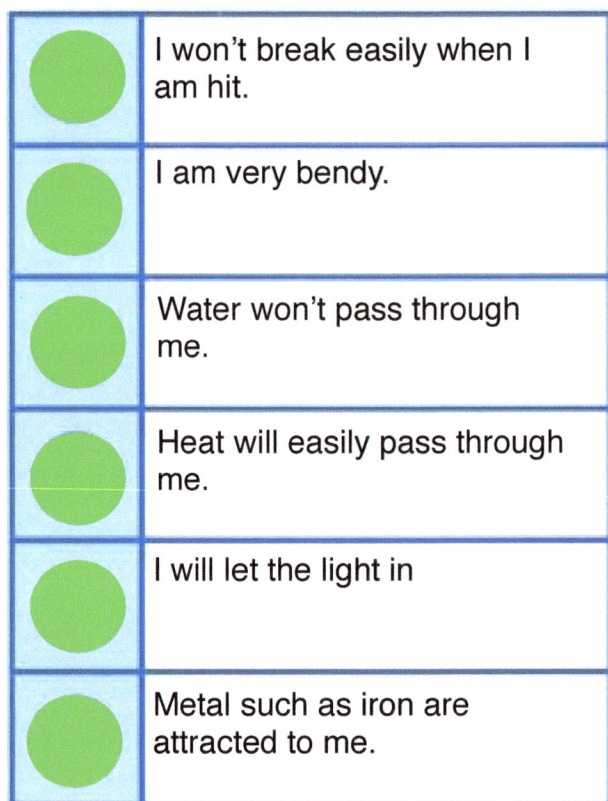

⬤	I won't break easily when I am hit.
⬤	I am very bendy.
⬤	Water won't pass through me.
⬤	Heat will easily pass through me.
⬤	I will let the light in
⬤	Metal such as iron are attracted to me.

Material Riddles.

I am flexible,
waterproof and strong.

I am flexible but not
strong or waterproof.

I am strong,
conductive and
magnetic.

I am strong and
conductive but not
magnetic.

I am strong and
waterproof but not
flexible or conductive.

I'm transparent and
waterproof but neither
flexible nor strong.

I am waterproof but
not strong or flexible.

Hints

Physical Properties of Materials Quiz

1. Magnetism is a physical property of a material. True or false?

2. _____ materials do not allow water to pass through them.
   - a. transparent
   - b. magnetic
   - c. water proof

3. _____ is a flexible material.
   - a. wood
   - b. rubber
   - c. plastic

4. Glass is a strong material. True or false?

5. Conductive materials allow _____ to pass through them.
   - a. light
   - b. heat
   - c. water

6. Iron is a magnetic material. True or false?

Newburyport, MA 01950

1-800-596-3175

OnBoard Academics employs teachers to make lessons for teachers! We create and
publish a wide range of aligned lessons in math, science and ELA for use on most
EdTech devices including whiteboard, tablets, computers and pdfs for printing.

All of our lessons are aligned to the common core, the Next Generation Science
Standards and all state standards.

If you like our products please visit our website for information on individual lessons,
teachers licenses, building licenses, district licenses and subscriptions.

Thank you for using OnBoard Academic products.